SMART GRIDS

and Other Energy Tech

Co-published by agreement between Shi Tu Hui and World Book, Inc.

Shi Tu Hui
Room 1807, Block 1,
#3 West Dawang Road
Chaoyang District, Beijing 100025
P.R. China

World Book, Inc
180 North LaSalle Street
Suite 900
Chicago, Illinois 60601
USA

Library of Congress Cataloging-in-Publication Data for this volume has been applied for.

Cool Tech (set #2)
ISBN: 978-0-7166-5387-5 (set, hc)

Smart Grids and Other Energy Tech
ISBN: 978-0-7166-5392-9 (hc)

Also available as:
ISBN: 978-0-7166-5398-1 (e-book)
ISBN: 978-0-7166-5404-9 (soft cover)

Written by Tom Jackson

STAFF

VP, Editorial: Tom Evans

Manager, New Product: Nicholas Kilzer

Curriculum Designer: Caroline Davidson

Proofreader: Nathalie Strassheim

Coordinator, Design Development & Production:
 Brenda Tropinski

Senior Media Editor: Rosalia Bledsoe

Developed with World Book by
White-Thomson Publishing LTD
www.wtpub.co.uk

ACKNOWLEDGMENTS

Cover   © Arturnichiporenko/Shutterstock
5-7   © Shutterstock
8-9   © Alessandro28/Shutterstock; © eye35.pix /
         Alamy Images; © Pi-Lens/Shutterstock; © SynthEx/
         Shutterstock; © Alessandro28/Shutterstock
10-11   © Shutterstock
12-13   © sakkmesterke/Shutterstock; © imageBROKER/Alamy
         Images; © metamorworks/Shutterstock
14-17   © Shutterstock
18-19   © Thinnapob Proongsak, Shutterstock; © Novikov
         Aleksey, Shutterstock; © Solar AquaGrid LLC; © Idpreya
         Del, Shutterstock
20-21   © Tom Buysse, Shutterstock; © Jacques Tarnero,
         Shutterstock; © David Bleeker Photography/Alamy
         Images; © DOE Photo/Alamy Images
22-23   © Breedfoto/Shutterstock; © emel82/Shutterstock; ©
         Evgeny V, Shutterstock; © Wavepiston
24-25   © RAW-films/Shutterstock; © Rudmer Zwerver,
         Shutterstock
26-27   © Shutterstock

28-29   © Akhmad Dody Firmansyah, Shutterstock; ©
         VectorMine/Shutterstock; © NuScale Power; © Bill
         Miller, Alamy Images
30-31   © guteksk7/Shutterstock; © Naeblys/Shutterstock; ©
         ZUMA Press/Alamy Images
32-33   © Shutterstock
34-35   © Val Thoermer, Alamy Images; © Have a nice day
         Photo/Shutterstock; © Jens Mommens, Shutterstock; ©
         petrmalinak/Shutterstock
36-37   © Manfred Dietsch, Alamy Images; © burakyalcin/
         Shutterstock; © VectorMine/Shutterstock
38-39   © Energy Vault
40-41   © Xinhua/Alamy Images; © Xinhua/Alamy Images; ©
         Kobus Smit, Shutterstock
42-43   © Audio und werbung/Shutterstock; © Vladi333/
         Shutterstock
44-45   © Iain Masterton, Alamy Images; © B Christopher,
         Alamy Images; © luchschen/Adobe Stock; © Alexander
         Limbach, Shutterstock

CONTENTS

Acknowledgments...................................2

Glossary 4

Introduction.......................................5

(1) Smart Grids...................................6

(2) Renewable Power 16

(3) Power Plants 24

(4) Storing Energy 32

(5) Hydrogen Fuel 42

Resources 46

Index..48

There is a glossary of terms on the first page. Terms defined in the glossary are in boldface type that **looks like this** on their first appearance on any spread (two facing pages).

GLOSSARY

fossil fuel a source of energy that formed from the remains of living things that died millions of years ago. Coal, oil, and natural gas are fossil fuels.

greenhouse gas a gas that warms Earth by trapping solar heat reflected from Earth's surface, much like the glass in a greenhouse.

hydroelectric any method that generates electricity from the power of flowing or falling water.

infrared an invisible portion of light that we feel as heat.

laser a device that produces a narrow and intense beam of light of only one wavelength going in one direction. The special qualities of laser light make it ideal for a variety of applications.

nuclear reactor a device that generates power using the energy released by changes in the nucleus (core) of atoms.

piston a disk or cylinder fitting closely within a tube in which it moves up and down against a liquid or gas. A piston is moved by a force, such as the pressure of steam or water, and transmits motion that can be used to generate power.

renewable energy from natural resources that can be used over and over. It includes energy from the sun, from wind, from moving water, and from heat beneath the ground.

solar furnace a device that uses concentrated solar (sun) energy to produce high temperatures.

solar panel a device that converts solar (sun) energy into electricity.

substation the equipment used to transform electric voltage from high to low or from low to high as necessary in an electrical grid.

tide a periodic motion of water in the oceans. Tides are caused by the gravitational pull of the moon and sun on Earth. Tides cause the water level at any place in the oceans to rise and fall in regular cycles.

transformer a device that transfers electricity to increase or decrease the voltage.

turbine a machine for producing power, in which a wheel or rotor is made to revolve by a fast-moving flow of air, gas, or water.

watt a unit in the metric system commonly used to measure electric power. The symbol for the watt is W. For household use, power is measured in kilowatts. One kilowatt equals 1,000 watts. For power plants, use is measured in megawatts (millions of watts) or gigawatts (billions of watts).

INTRODUCTION

Modern cities rely on a steady supply of abundant electricity. And the demand for electricity is growing fast. The electrical supply system of a city is called the grid. It is an enormous network of crisscrossing wires and cables that stretches from a power plant to your desk—and it is almost impossible to imagine life without it. But today's grid is not much different from when it was first developed 130 years ago. Power outages occur more frequently as the old grid strains under the demand. This old-fashioned energy technology cannot possibly meet the demands of the future.

The old technology used to generate and distribute electricity today causes other problems, too. It is dirty, creates pollution, and is a major contributor to climate change. Traditional sources of power to generate electricity will run out one day. What will we use then?

The cities of the future will need a complete overhaul of energy grid technology. This new grid will have to distribute electricity sensibly and not waste energy or increase pollution. It will rely on information about where electricity is generated and where it is used every second. The electrical grids of the future will use artificial intelligence (AI) and other advanced technology to control the electricity supply. Sources of **renewable** energy will replace the old fuels to generate electricity. These sources will have to be cleaner and safer than the old energy technology. Households may even become generators. In short, the cities of the future will need smart grids and other energy tech.

The smart grid of the future will provide cleaner electricity generated efficiently from many different sources, such as solar farms, **wind turbines,** geothermal plants, or perhaps nuclear fusion reactors and hydrogen. Smart meters will monitor how, when, and where electricity demand is greatest and react accordingly. When the demand is lower, a smart grid will store electricity for future use. Power plants that generate electricity will be smaller and cleaner.

1 SMART GRIDS

GETTING SMARTER

If tomorrow's energy grid is to be smart, today's grid is still quite dumb. A modern electrical grid is designed to carry electricity in one direction only. It goes from the power plant where it is generated to the homes and factories where it is used. It never goes the other way.

In an old-fashioned electrical grid, electricity is produced in huge power plants—each one big enough to supply an entire city. Even today, most power plants generate electricity by burning **fossil fuels** as they did a century ago. Burning fossil fuels produces dangerous pollution and waste gases that contribute to climate change.

Engineers have figured out cleaner and safer ways to generate electricity, such as solar and wind power. These technologies do away with enormous power plants. Unlike fossil fuels, they are renewable. Solar and wind generate clean electricity from a network of small generating systems spread out across the land and the ocean. But the source of that power—the sun and wind—is not as steady and reliable as fossil fuels. The network of connections to transport and store that power is also more complicated.

So what's the holdup? Today's dumb electrical grid has difficulty handling variable sources of power like wind and solar energy—the fastest-growing sources of renewable power. The best places to generate wind and solar power are located in remote places. But most demand for electricity is in large, dense cities.

A new smart grid is needed to manage the flow of electricity from different suppliers to users. The supply has to be monitored and sometimes stored. Some of those users will produce energy, too. The smart grid is a new technology that will replace the old energy technology in the near future. It will take a while to replace the dumb grid completely. But some of that new smart technology is already in use.

SMART METER

The smart grids that will power the cities of the future will require changes in the way electricity is generated and used. A consumer uses electricity. That could be in your home, an electric car, or a big office building or factory. A producer is a power plant on a grid that is generating electricity. Above all else, a smart grid needs information about how much electricity is used at different places at different times. One way to collect that information is by using smart meters.

Smart meter. The smart meter is a measuring device fitted in all homes and on a smart grid. The meters record how much electricity is used at any given moment. The information is sent to a control center through the Internet or by a text message, and the power plant can adjust its operations to meet demand as it rises and drops throughout a day, week, month, or year.

Two-way communication. The digital technology that allows this two-way communication between the power plant and customers is what makes the grid smart. At its simplest, a smart meter measures how much electricity a consumer uses at various times. The power company then knows to charge them accordingly for the service. Consumers can save money and reduce electricity demand at peak times by switching off unnecessary devices.

The smart grid also allows energy suppliers to respond digitally to the quickly changing electricity demand of cities. A heat wave may cause a sudden surge in electricity demand as consumers turn up their air conditioners. A smart grid will alert a power plant that it needs to increase the output of electricity. The grid will also send a message to customers via email or text instructing people to turn off devices to reduce strain on the electrical system. This helps prevent outages.

| VOLTAGE | 222.09V | CURRENT | 18.1878A |
| POWER | 4039.33W | PF | 0.6933 |

⚠ **POWER GRID FAILURE!**

793.3104kWh	13.2225kWh
$305.58	$2.86
29d	1d

Home supply. On a smart grid, consumers can also become electricity producers. On sunny days, a home with **solar panels** may generate more electricity than it uses. The smart grid allows that power to flow from the house to other consumers. A smart meter records the surplus electricity and credits the account. This way, everybody benefits, and no valuable energy is wasted.

DECENTRALIZED POWER

An important feature of new smart grids is that electricity production is *decentralized*. That means electricity is generated in many different ways in different places. Instead of a single large power plant, practically every home on a block can be a mini power plant. They may generate electricity through solar panels or wind turbines. Such spread-out generation systems are called *distributed energy resources* (DER's). Managing DER's to supply electricity in a big city is difficult. But smart grid technology is being developed to get it right.

On and off supplies. A smart grid will connect all the mini power-generating resources through a system of smart meters and controls. On cloudy days, solar panels generate little electricity. Wind turbines stop producing power if the wind speed drops. When this happens, the smart grid can immediately switch to electricity producers that do not rely on the weather. This could be a nuclear power plant or a **hydroelectric** dam, or a power plant that generates electricity by burning fossil fuels. The smart grid will switch on a dirty power plant if consumer electricity demand rises and there are not enough clean DER's to supply it.

Voltage. A smart grid made up of DER's must also control voltage—a measure of the force pushing electricity through the wires. Power plants send electrical current over wires at a very high voltage, but homes use a much lower, safer voltage. Typically, a **transformer** increases or reduces voltage for the electrical supply between a power plant and a home. In a smart grid, a transformer must control voltage moving back and forth between DER's, homes, and power plants.

Islanding. During an outage, the smart grid will allow others in your community to grab electricity from your solar panels—and your neighbor's—to keep the lights on even when there is no power coming from a power plant. This is called *islanding.* It allows homes on a smart grid to keep the lights on from DER's until workers can bring the main power plant back online.

AI AND THE SMART GRID

The hardware of a smart grid includes the wires that carry electricity, the **substations** that direct the flow of electricity from generators, and the smart meters that monitor electricity usage. Together, this requires millions of lines of software (computer programs) that control all the hardware. To manage this incredibly complex system, the smart grid uses artificial intelligence (AI). AI is a computer system that processes information and learns more like humans think.

An expert system. Human experts have programmed smart grid AI to make decisions and keep the electricity flowing. For example, the smart grid can be programmed to receive weather forecasts. Smart grid AI uses the forecasts to predict how much electricity will be produced by solar panels and wind turbines in a DER. It can then instruct a power plant to increase or decrease its output accordingly.

Machine learning. Smart grid AI may also use *machine learning*—where the AI teaches itself to see activity patterns in the grid. Smart grid AI might record that consumers use more electricity in the morning but consumption drops during the day as people go to work. The AI may see that consumption rises again in the evening as people return home. Smart grid AI can also see activity patterns that are not so obvious to human controllers. Combing through usage data from thousands or even millions of consumers, AI can predict electricity supply and demand with far greater accuracy than any human. That means the smart grid can be controlled precisely without wasting energy.

Optimization. Once the AI learns the patterns of electricity usage in a grid, it calculates the best way to supply that electricity. This is called *optimization*. It may involve changes to only a portion of a smart grid at specific times. For example, a smart grid providing electricity for heating and cooling in Suzhou, China, used AI optimization to quickly show the best way to distribute electricity to maximize efficiency.

REDUCING WASTE AND DISRUPTION

Smart grid technology aims to make using electricity more efficient. That means not wasting electricity. DER's may generate great amounts of electricity even when demand is low. With old dumb grid technology, surplus electricity is usually wasted. With smart grid technology, the excess electricity can be stored for use later.

Virtual power plants. The electrical storage systems of a smart grid could be made to work together to supply power just like a power plant does. This idea is called a virtual power plant (VPP). In 2022, Tesla partnered with the California power generator PG&E to have consumers store surplus electricity in a household battery named Powerwall. When PG&E power plants struggle to meet the high demand that occurs during heat waves, homeowners can send any surplus electricity back through the smart grid to help ease the strain. A network of thousands of Powerwalls in a region can serve as a huge distributed battery—a virtual power plant.

VPP's can also prevent power outages during heat waves and other instances of high demand. They can also take over the power supply for a limited area where power plants or power cables are damaged by extreme weather or accidents.

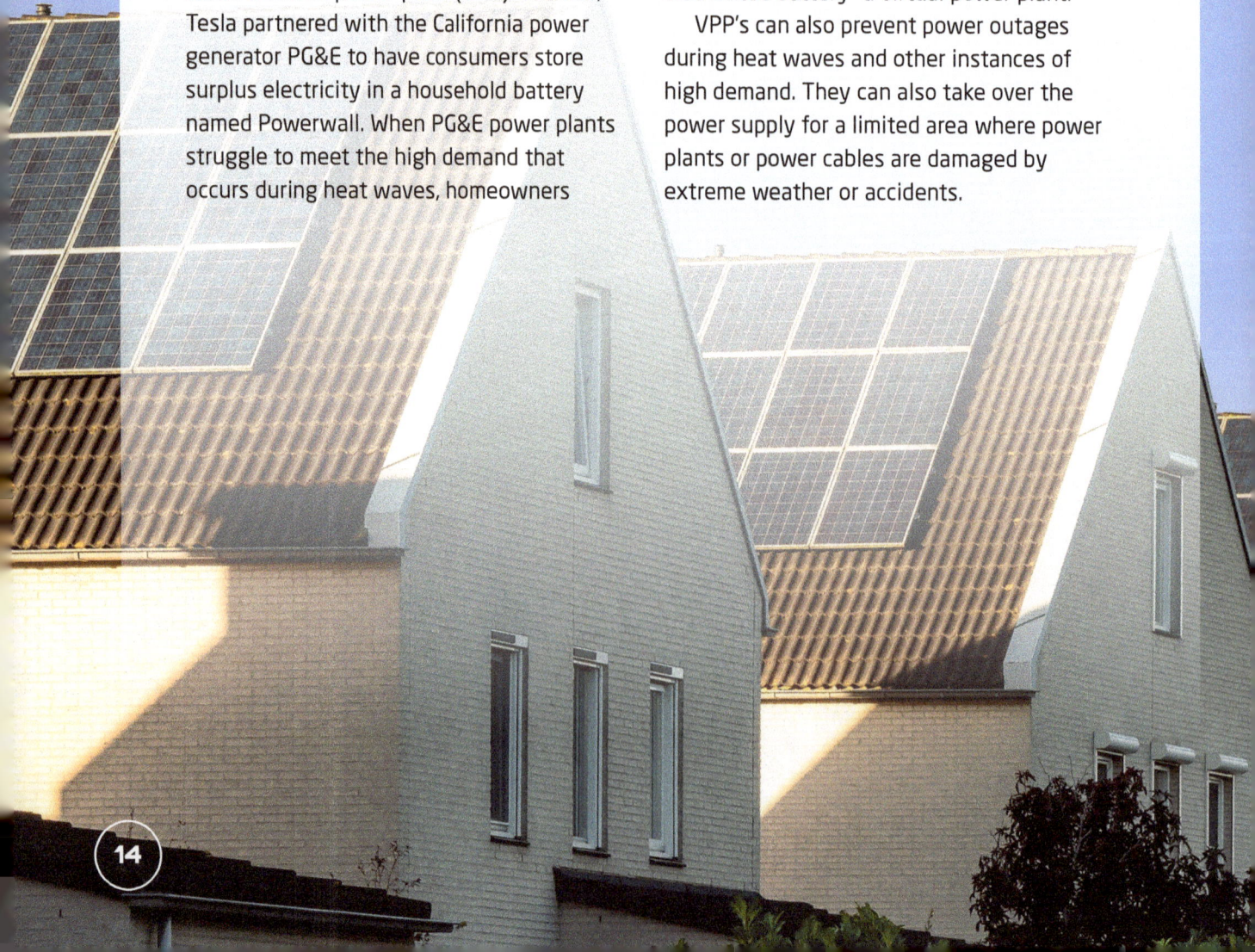

Home area network. Smart grid technology is not limited to city-wide electricity distribution. Individual houses and apartment buildings can have their own mini smart grid, known as a home area network (HAN). In a smart grid home, televisions, washing machines, computers, and other devices are the consumers. The HAN controls when these appliances are switched on. For example, washing machines could run at night when electricity is less expensive.

A smart grid home may produce its own electricity, perhaps with solar panels on the roof. The HAN uses this power to run the house. The HAN will only draw electricity from the main grid when this home supply runs low. When a home has surplus electricity, the HAN directs it for storage in a rechargeable battery for the home or an electric vehicle (EV) parked in the garage. Those same batteries can be used by the HAN to power household appliances if needed.

2 RENEWABLE POWER

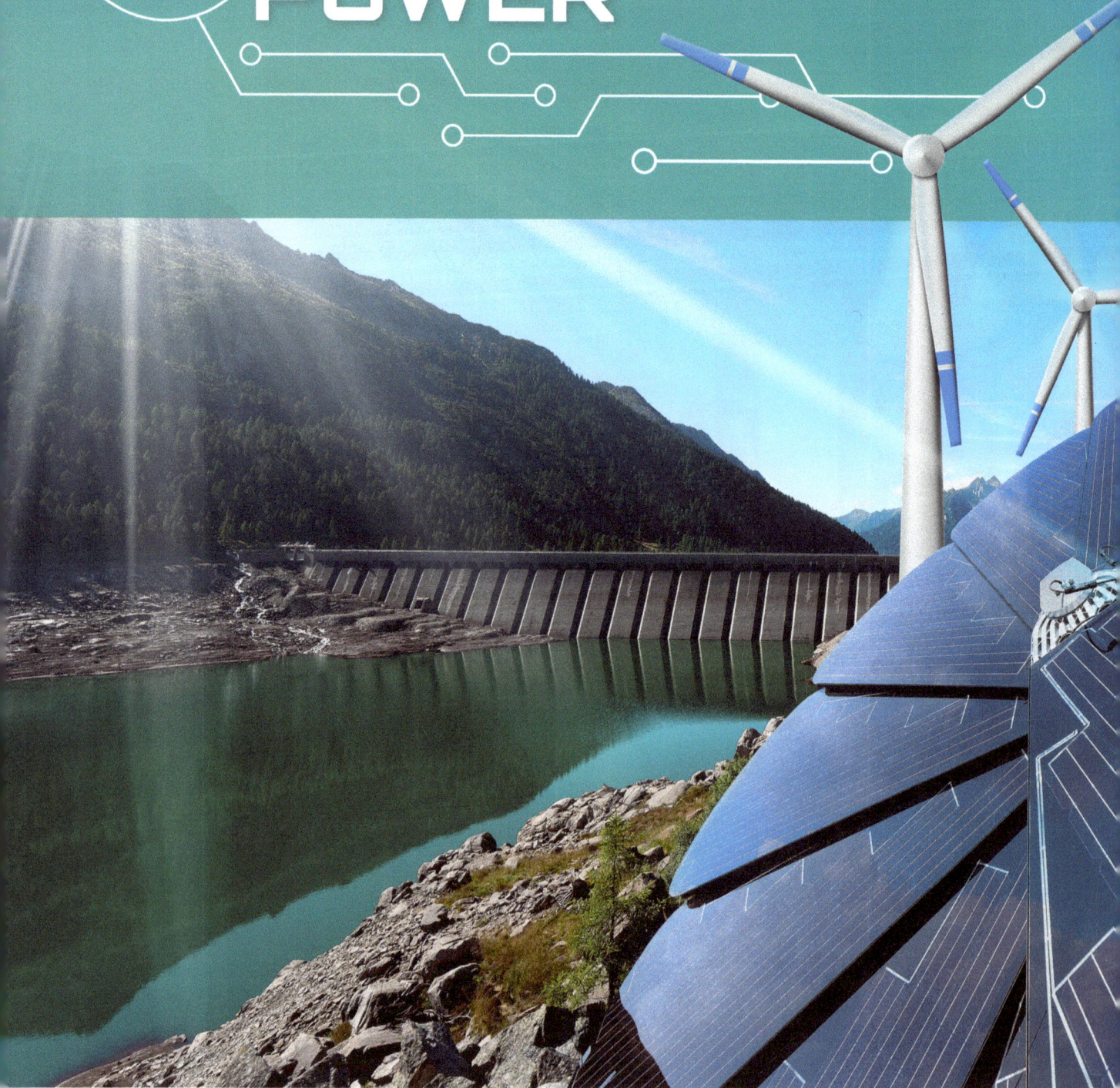

CLEAN BUT UNRELIABLE

Smart grid technology is ideal for the renewable energy generators that will power the cities of the future. Renewable power sources include solar power, wind power, geothermal energy, and tidal power. Renewable power sources never run out no matter how much electricity they generate, and they do not contribute to climate change

But sources of renewable energy vary, and it is impossible to control them. On some days, renewable power sources may generate much more electricity than we need. On other days, they may generate almost nothing. This is where smart grid technology can help. Smart grids smooth out the supply through DER's and save extra power for when it is needed. But to decrease the reliance on fossil fuels, renewable power tech must improve, so that we can harness this never-ending supply of energy to power the cities of the future.

THE POWER OF THE SUN

The amount of energy carried by the sunlight shining on Earth for just one hour is more than enough to meet all of our energy needs. Solar power is a potentially limitless source of clean renewable energy. Solar power used to generate electricity is growing fast in many parts of the world. But solar power makes up only about 4 percent of the electricity generated today. To power the smart grids in cities of the future, solar power generation will have to be scaled up. What kinds of cool new tech can increase solar power?

Thermophotovoltaics. Traditional solar panels generate electricity through the photovoltaic effect. These use light energy to generate electricity, and they don't work well on cloudy days. But most of the energy from the sun is invisible **infrared,** which we feel as heat on our skin. New thermophotovoltaic panels generate electricity from heat as well as light.

Thermophotovoltaic panels work on cloudy days and even pick up heat rising from the ground at night.

Solar canals. Solar power works best in sunny parts of the world—places that also tend to be hot and dry, like Los Angeles, California. This huge desert city has needs, too, namely water. In 2023, city engineers began a project to construct solar panel arrays over the canals that transport water to Los Angeles. The plan is to cover the canals to prevent water from evaporating in the hot sun. At the same time, the panels will generate renewable energy as the city tries to reduce its dependence on fossil fuels for electricity.

Engineers think that solar panels placed over the 4,000 miles (6,437 kilometers) of California's canals could save 65 billion gallons (246 billion liters) of water from evaporation every year. The solar panels will also generate 13 gigawatts (13 billion **watts)** of renewable electricity to power millions of homes.

Power tower. Concentrated power uses curved mirrors to focus light into an intensely bright and hot beam. That beam can be used to heat a liquid inside a central column called a power tower. The liquid is melted salt, which is good at holding on to heat and can be used in place of fuels for generating electricity. The heat of the beam can also be used as a **solar furnace** for melting metals and other industrial uses.

Artificial photosynthesis. A leaf is a natural solar panel that collects energy from the sun and turns it into energy to power life processes and growth. This process is called *photosynthesis*. But plants are only so good at turning light into energy. Plants waste about half the energy in light. Today, scientists are hacking this natural process. They are finding ways to make the chemical reactions of photosynthesis more efficient and use light to produce liquid fuel.

Efficient photosynthesis would use all the sunlight's energy—including the green light that is normally reflected. So leaves would look black!

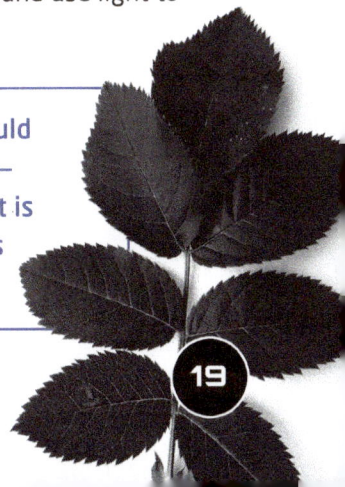

WIND POWER

Wind turbines capture the motion of the air with rotor blades. The spinning rotors drive a generator to make clean electricity. There is no burning of fuel that creates pollution or **greenhouse gases** that contribute to climate change. Humans have used the wind to power devices for centuries—think of windmills. It is not exactly a new technology! But wind power is perfect for the smart grids that will power our future. So researchers are developing amazing new designs to improve this well-known technology.

Giant turbine. Wind turbines can be built anywhere it is windy. The best places for turbines are out at sea, where there are no hills or valleys to block the wind. Offshore turbines can also be made much larger than those on land. In 2023, the China State Shipbuilding Corporation (CSSC) announced that it will construct the largest and most powerful wind turbines ever. Designed for offshore use, the three-bladed turbines will reach an enormous 853 feet (260 meters) in diameter. As they spin in the wind. the long rotor blades will sweep over an area equivalent to nine soccer fields. Just one turn of the blades will generate enough electricity to power a house for two days!

Deep water. Large offshore wind farms already exist along many coasts where the water is shallow enough to easily build the turbines. But vast amounts of wind power can be generated farther offshore in deeper water. To capture more of the ocean's wind, researchers are designing turbines to float in deep water. These new turbines are designed so that most of the structure is underwater: This helps it stay upright during storms.

Look, no blades! Wind turbines have a lot of moving parts. If anything breaks, the whole turbine stops working. Electrostatic Wind Energy Converters (EWICON) don't have that problem—because they have no spinning blades or moving parts. This new wind power generator sprays electrically charged water droplets from a grid. The grid itself carries the opposite charge, so the droplets are pulled back to the EWICON by electrical forces. However, as wind blows the droplets away, that same electrical force pulls a flow of charge from the ground, creating an electrical current.

WATER POWER

Earth's moving water is a vast source of renewable energy, most of which is yet to be tapped. Today, about half of the world's renewable power is produced by hydroelectric dams that generate electricity from the motion of falling water. But researchers are inventing new technology to generate more electricity from water in different ways to power the smart grid.

Earth's **tides** make the ocean rise and fall. The flow of the tides is predictable and everlasting—a perfect source of clean renewable energy if it can be tapped. But it is difficult to build a power plant in the ocean.

Tidal lagoon. Researchers have developed technology to generate electricity from the movements of Earth's tides. One cool new design is called a tidal lagoon. To construct a tidal lagoon, engineers build a vast basin surrounded by a wall on the coast. Water flowing with the rising tide flows through turbines within the wall to generate electricity. As the tide rises, the lagoon is filled as water rushes through the turbines, generating electricity.

At high tide, the lagoon is sealed, so the water stays inside as the tide falls. Later, the water is allowed to drain back into the ocean, driving the turbines and generating power. The tides rise and fall twice each day, so tidal lagoons can generate clean renewable electricity continuously.

Wave power. Ocean waves are a potential source of enormous amounts of everlasting water power. But generating electricity from ocean waves presents many technological challenges. Sea Wave Energy Limited (SWEL) is developing technology that captures energy from ocean waves to generate electricity. In 2022, they launched their prototype device, the Waveline Magnet—a linked group of floating platforms made from recycled plastics that bends and flexes on ocean waves. The motion is used to generate electricity. The company expects a single unit to generate 100 megawatts of electricity at a cost comparable to that of using current fossil fuels.

New wave power systems also include wave pistons—hollow columns placed offshore. Ocean waves push water in and out of the column, driving a **piston** inside that generates electricity. The Danish company Wavepiston uses a linked chain of energy collectors anchored in the ocean. As waves pass, they move plates back and forth. This motion pumps seawater that flows past a turbine, generating electricity. The experimental technology may be used to generate electricity for smart grids in coastal regions.

3 POWER PLANTS

LARGE-SCALE POWER

A smart grid will still need large power plants to create the *base load.* This is the minimum amount of electricity necessary for the grid to operate. If the electricity supply falls below the base load, power outages will occur in the network. Modern electrical grids need a reliable energy producer to ensure that the base load is met in all conditions. Today's large power plants may be dirty and inefficient, but they are reliable.

The smart grids of the future will use more clean and renewable power sources to generate electricity. However, these sources are currently not as reliable as they need to be. New technology will be necessary to increase levels of clean renewable power to reliably meet the base load demand. Other new technology can help existing dirty power plants meet base load demand with less pollution and greenhouse gas generation.

Today's dirty power plants are used as *peakers.* That means that they are turned on when electricity demand is highest, or at its peak. In a smart grid, this peak demand will be met using stored electricity. But older power plants may still be necessary to meet the highest demands for electricity. New technology can help these old plants operate more cleanly and efficiently.

USING HEAT

Today's coal and gas-fueled power plants generate heat (thermal energy) used to boil water into steam that spins turbines to generate electricity. This thermal power system works well on a large scale. But it also generates pollution and greenhouse gases that contribute to climate change. Older power plants used to meet base load and peak demand can be improved with new technology, so they can operate cleanly.

Garbage power. Amager Slope in Copenhagen, Denmark, is a vast thermal power plant that burns garbage to generate electricity. Any waste that cannot be recycled can be used as fuel, so it helps reduce garbage pollution. The plant uses filtering *scrubbers* to remove dirt and harmful substances released when garbage is burned—so it is cleaner than older thermal power plants.

Scrubber technology works in several ways. Smoke from burning fuel is sprayed with chemicals that react with polluting substances in the smoke. The smoke then passes through a powerful electric field before it goes up the smokestack. The electric field pulls dust and soot from the smoke, leaving behind only clean gases.

Carbon capture. Garbage can be burned as a clean, nonpolluting fuel in power plants equipped with scrubbing technology. But the burning still releases carbon dioxide (CO_2), a major greenhouse gas that contributes to climate change. Power plants today rely on carbon-capture technology to remove CO_2 from the smoke of burning fuel.

Today, most thermal power plants use aluminum formate (ALF) for carbon capture. ALF crystals are full of tiny holes. When the smoke from the power plant is pushed through an ALF filter, the carbon dioxide gas is trapped in the holes and not released into the atmosphere.

Data centers. Heat to power a thermal plant does not have to come from burning fuel. It is produced by machines, such as computers. Today's data centers—warehouses filled with the computers that make up the internet—produce great amounts of waste heat from the electrical components. Researchers have figured out ways to recycle that heat to warm a swimming pool, heat a building, and even generate modest amounts of electricity that can return to the smart grid. The Internet uses about 5 percent of the world's electricity, so there are a lot of hot computers out there!

GEOTHERMAL AND NUCLEAR POWER

Earth is very hot inside. Every 300 feet (100 meters) beneath the surface, the temperature rises by about 5 °F (3 °C). The heat is produced by natural **nuclear reactions** deep inside Earth. Some modern power plants make use of Earth's heat and the same nuclear reactions that create it to generate electricity. New technology is making these power plants smaller and safer.

Geothermal power. The heat of Earth is called geothermal energy. A geothermal power plant operates much the same way as any thermal plant. Instead of burning fuel to generate high-pressure steam, a geothermal plant pumps water into the earth, where it is heated by hot rock. Today's geothermal plants are built in volcanic regions. Here, the hot rocks and water are near the surface for the power plant to tap into.

New enhanced geothermal technology creates artificial *hydrothermal* (hot earth and water) systems by pumping water into the hot, dry rock beneath Earth's surface. In the future, these geothermal plants could be built almost anywhere—they just need to be deep enough. The water must be pumped into hot rock layers about 5 miles (8 kilometers) beneath the surface for the system to work.

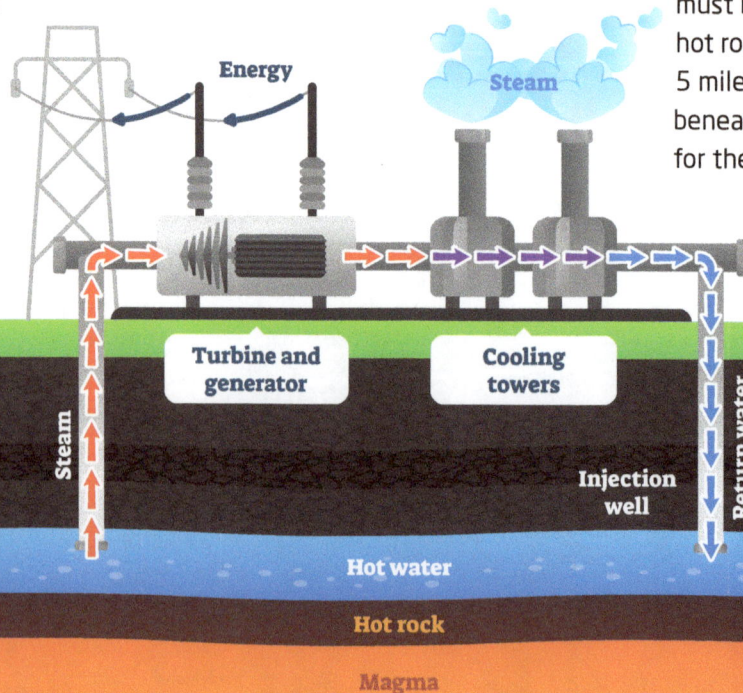

Energy

Steam

Turbine and generator

Cooling towers

Steam

Injection well

Return water

Hot water

Hot rock

Magma

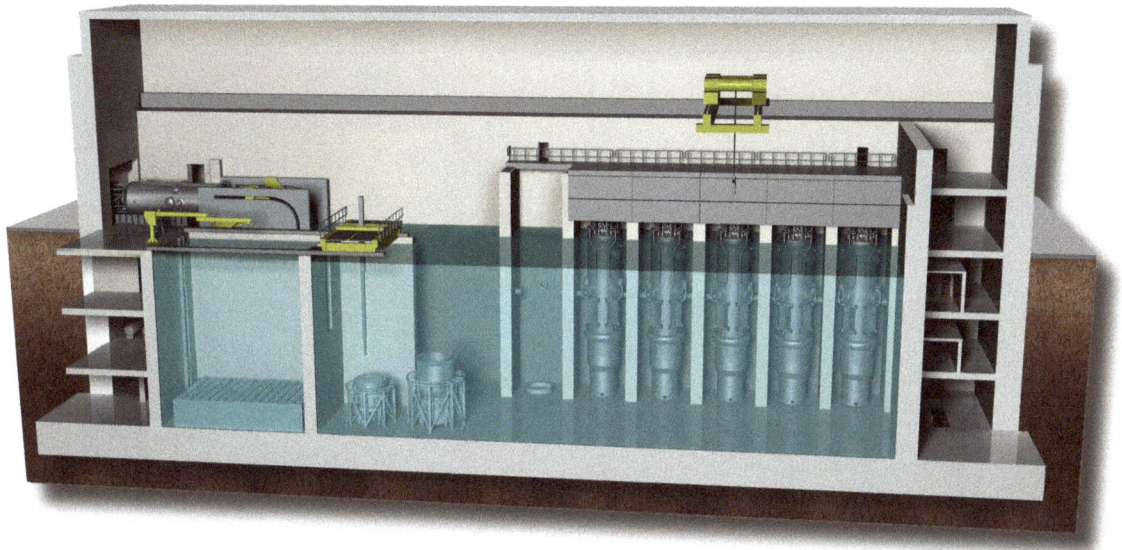

Small reactors. **Nuclear reactors** use the heat from nuclear reactions to generate electricity. Nuclear power plants are among the largest in the world, and they take many years to construct. Engineers are designing new nuclear power systems that are smaller and easier to construct. Small nuclear plants could be used to provide base load electricity to smart grids in remote places where solar or wind power is not feasible.

These new reactors will be *modular*—constructed in similar sizes with similar units for flexibility and variety in use—and portable. They can be mass-produced and transported to where they are needed. The small reactors come loaded with fuel and can operate for years without refueling.

Safer nuclear. One danger from nuclear reactors is a serious malfunction known as a meltdown. This occurs when nuclear fuel overheats and melts. Chemical reactions between the nuclear fuel and the water used to cool the fuel release explosive hydrogen gas. New reactor designs use liquid sodium or molten salt for cooling. These liquids are safer at high temperatures and do not release dangerous hydrogen when things go wrong.

A sodium-cooled reactor system is being tested at Monju in Japan.

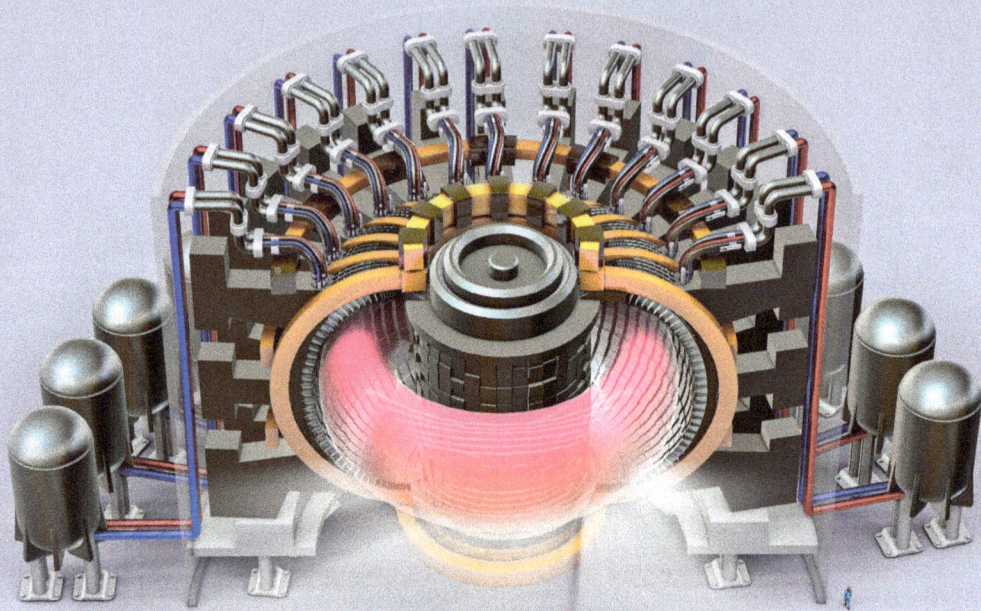

NUCLEAR FUSION

Nuclear power plants of the future will be more powerful and far less dangerous than those used today. Today's nuclear power technology generates heat through *atomic fission*—splitting atoms. Scientists are working to design new reactors that produce heat by *nuclear fusion*—where two atoms are combined. Nuclear fusion is the sun's power source. To construct a nuclear fusion reactor, engineers must recreate the conditions inside a star. That's a real challenge! A practical nuclear fusion reactor has yet to be invented. But scientists are working to develop nuclear fusion technology because it promises an unlimited supply of safe, clean energy.

Tokamak. Most researchers are working to develop a nuclear fusion reactor using a design called a tokamak. This is a doughnut-shaped chamber that energizes hydrogen atoms into a superheated, electrified form of matter called *plasma*. Powerful magnets hold the plasma in streams that whiz around the reactor. When they are moving fast enough, the energized hydrogen atoms smash together and fuse. This releases tremendous amounts of energy. Today, creating the plasma streams uses more energy than it produces—it does not generate new power.

ITER. The world's largest tokamak, the ITER (Latin for "the way"), is under construction in France. Researchers believe that this experimental reactor will be powerful enough to generate a steady fusion reaction. This will allow them to study fusion reactions, but ITER is not designed to generate electricity.

Ignition achieved. Experimental fusion reactors today are unable to produce a useful supply of energy. To do that, a fusion reactor must produce more energy than it uses, a milestone achievement called *ignition*. In 2022, researchers at the Lawrence Livermore National Laboratory in California used a different method to start a fusion reaction. This system used 192 **lasers** to heat a diamond capsule of hydrogen atoms until the atoms fused. The reaction briefly released more energy than it used. This was the first time scientists achieved nuclear fusion ignition—a major goal. However, nuclear fusion technology remains a long way from generating useful amounts of energy or electricity.

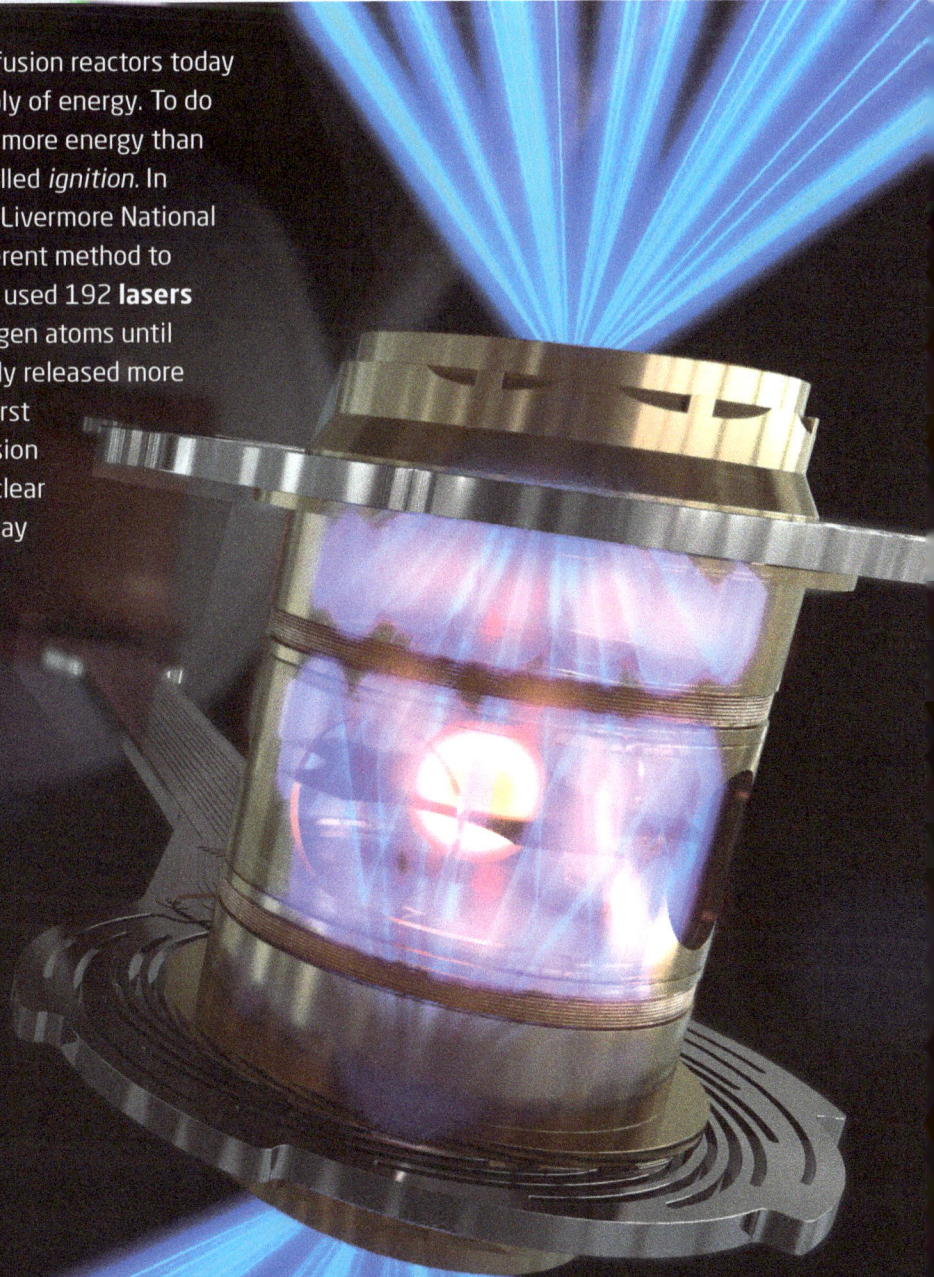

4 STORING ENERGY

SOLAR ENERGY STORAGE

ENERGY STORAGE

WASTE NOT, WANT NOT

A future smart grid system must ensure a reliable supply of electricity for everyone. Rather than powering-up to meet increasing demand, the smart grid taps into stored electricity to increase supply. Today's dumb electrical grids have little electrical storage capacity. That will soon change as the world moves to the power sources of the smart grid.

Electricity is not easy to store. Today, electric current is often collected in rechargeable batteries. But these can only hold the current for a limited time. The current steadily leaks out of the batteries over time, especially if they are heated. One method to store power longer is to convert the electrical energy into another kind of energy.

New energy storage technology is used to make better batteries. It is also using novel methods to store electrical energy in other forms—including gravitational energy, compressed air systems, and pumped storage hydropower. These new technologies use fairly simple concepts to develop innovative new energy storage systems.

BATTERIES

A battery stores chemical energy and releases it as electricity. Smart grid battery storage systems use rechargeable batteries—where electricity is input and stored as chemical energy for later use. Rechargeable batteries are an important component in modern electrical vehicles (EV's). These are improving rapidly as well. Soon, your EV battery may even have a role in the smart grid!

Parked cars. Any large rechargeable battery found in an EV can also be used to store electricity. An EV is generally plugged into the grid to recharge when it is not in use. A smart grid can use any parked EV's as a VPP (virtual power plant) when supplies from other DER's are running low.

On the move. Modern EV's use bulky lithium-ion batteries that weigh 1,000 to 2,000 pounds (454 to 900 kilograms). The EV of the future will need a smaller and lighter battery that increases the EV range (distance it can travel before recharging). Engineers are designing a new generation of batteries that use nickel and graphite (a type of carbon), sulfur, and even air to hold and release electric current. These new batteries will charge much faster, too. Today, the fastest EV chargers are used with Formula E electric racing cars. The charger can deliver enough power for 25 minutes of high-speed racing in just 30 seconds. A system like that could fully recharge a regular EV in about 15 minutes!

Sodium batteries. Lithium is a lightweight metal used to make the lithium-ion batteries in most EV's today. The high demand for lithium to power the growing EV market makes this material expensive. Sodium is a common metal with similar properties. Current sodium batteries do not work as well as lithium batteries—they are also heavier, but they are cheaper to produce. Sodium batteries are compact enough to be used for storing electricity in homes on the smart grid.

Big iron. Storing electricity does not always require small or lightweight batteries. Big storage batteries can be made from iron, which is cheap compared to lithium. However, iron batteries are inefficient—they lose about a fifth of their stored power as heat. But the demand for electrical storage in the smart grid is enormous. Large iron batteries are still useful for electricity storage to economically meet the demands of the smart grid.

ENERGY STORAGE

Lithium ion **battery system**

ENERGY STORAGE

Lithium ion **battery system**

PUMPED STORAGE HYDROPOWER

What goes up must come down. Today, an old renewable energy technology is being renewed as an efficient way to store energy for the smart grid. Hydroelectric (or hydropower) systems generate electricity using the downhill flow of water. A hydroelectric dam generates electricity from water naturally flowing through a river system. In contrast, pumped storage hydropower uses some electricity to pump water uphill. When more power is needed, the water is allowed to flow back downhill, turning its stored energy into electricity.

Pumped storage hydropower can never generate more electricity than it uses to pump the water uphill. That is impossible under the laws of physics! But the systems are useful in a smart grid for use when hydroelectric demand rises. This helps make the smart grid system more reliable.

Two reservoirs. A pumped storage hydropower facility has two reservoirs—one higher than the other. Water is pumped from the lower reservoir to the upper one. The water in the top reservoir has gravitational potential energy, which is converted into electrical energy when needed by flowing down through a turbine.

Open or closed. A closed loop pumped storage system uses a fixed amount of water transferred between the two reservoirs. Closed loop systems are often built on steep mountain slopes, so the upper reservoir is far above the lower one. An open loop system uses a river as the source of water for its lower reservoir. An open loop pumped storage facility is built next to a dam and shares its reservoir. The upper reservoir may not be as high, but the total water supply is larger than that in a closed loop system.

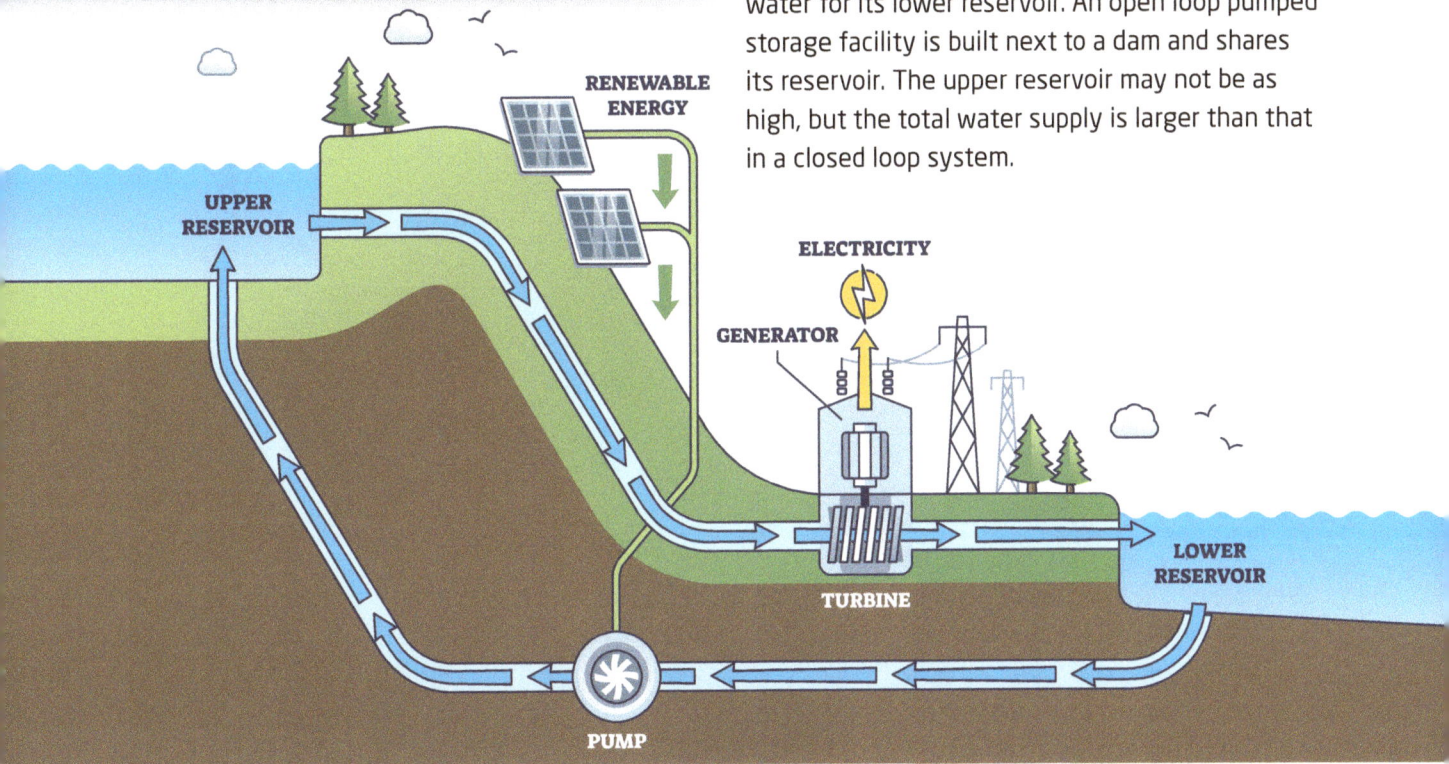

High density. Small-scale pumped storage systems can be built in places that do not have steep hills and large rivers. Instead of water, the pumps push a thick, heavy liquid up a slope. The system uses an artificial mud containing tiny grains of chalk or clay. The slippery heavy mud holds more gravitational potential energy when it is lifted. Such systems use smaller amounts of liquid not raised as high to create a modest DER on a hill.

GRAVITY BATTERY

Gravity battery technology is very new and needs a lot of testing. But this cool energy storage technology is ideal for a smart grid because it can be constructed almost anywhere. You may soon see gravity batteries next to solar and wind power plants in the smart grid of the future. A gravity battery stores energy using the same method that pumped storage hydropower uses: gravity. Instead of liquid, a gravity battery system lifts heavy, solid weights. Like pumped storage, a gravity battery cannot generate new energy. It can only store energy for use when it is needed.

Long drop. Modern gravity batteries have two main designs. One looks like a half-built skyscraper, with tall cranes that hoist heavy concrete blocks that are stacked like bricks into towers. Another design makes use of pre-exisiting mine shafts. Heavy blocks are moved up and down inside these deep holes. Both systems lift heavy blocks and drop them to release the stored gravitational energy.

Regenerative braking. A gravity battery relies on a concept known as regenerative braking. This technology is widely used in modern EV's. Brakes squeeze the EV wheels to slow the vehicle, creating *friction* (resistance) in the wheel. Regenerative brakes use this friction to generate electricity. That electricity is used to recharge the EV battery.

In a gravity battery, a pulley hoists a weight using the power from an electric motor. When needed, the weight is dropped. A regenerative braking system slows the weight's fall to create friction and generate electricity.

Rail energy. Another gravity battery design runs weights along a railroad. Advanced Rail Energy Storage (ARES) uses *mass cars* (weights) fitted with wheels and pulled up a steep slope on rails. Some of the energy used to pull the weight is stored as gravitational potential energy that can be used later. To release that energy, the mass cars roll downhill at a high speed. Regenerative brakes slow the mass car and generate electrical power from the friction.

COMPRESSED AIR SYSTEMS

Ordinary air can be used to store energy for the smart grid. Air is gas—a mixture of invisible atoms constantly bouncing around in all directions. Moving air particles hit objects, creating a force called pressure. When air is squeezed, or compressed, its particles fill a smaller space. Packed in tightly, air particles strike the walls of a container more often, and the pressure of the gas increases. Now, high-pressure air is used to store and generate electricity when needed.

Under pressure. Compressed air energy storage (CAES) systems work by squeezing air into a container. After the air is compressed (squeezed), each particle of air gains energy. When the gas is allowed to expand, the particles spread out from each other again. That spreading out creates a flow of fast-moving air that's used to drive a turbine and make electricity. The systems vary in size. Some CAES facilities keep the compressed gas in metal tanks about the size of a truck. Other facilities pump the high-pressure air into vast underground caverns.

Recycled heat. As the gases in a CAES system are squeezed, they become hotter. When the gases expand, they lose that heat and cool. A refrigerator and air conditioner make things cold by making gas expand. Most CAES systems take the heat from the high-pressure stored air and use it to warm up the gases when they are expanding. Adding this heat into the expanding gas increases the airflow and generates more electricity.

Ice battery. About a tenth of the world's electricity supply is used to keep us cool through air conditioning. The ice battery is a new technology used to make cooling systems more energy efficient. It is simple enough. At night, when the electricity demand is low on the grid, an ice battery system makes ice through refrigeration. In the daytime, when it is warmer, the ice cools a room without using electrically powered air conditioners.

5 HYDROGEN FUEL

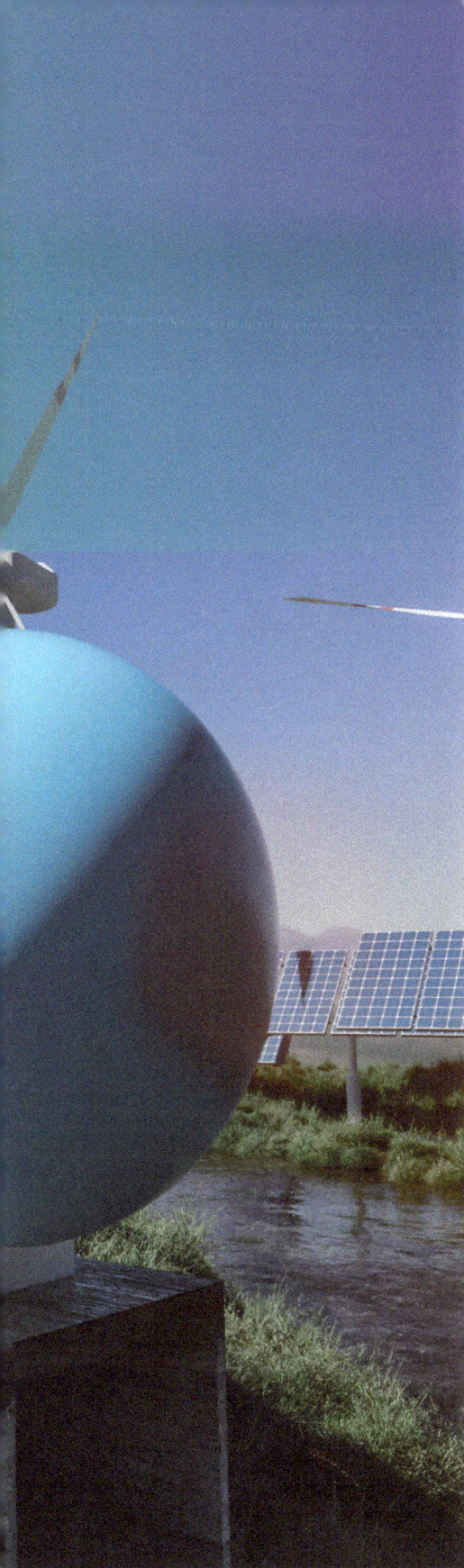

THE BEST FUEL EVER?

Smart grids are coming online and efficiently generating more electricity every year. But some machines cannot run on the electricity delivered by a smart grid. For example, electricity can only power small aircraft traveling short routes. Modern passenger jets require fuel packed full of energy—fossil fuels such as gasoline and kerosene (which is used to power jet aircraft). Fossil fuels used in global air transportation are major sources of harmful greenhouse gases. Is there an alternative fuel?

Hydrogen promises to be the fuel of the future. Hydrogen is the most abundant gas in the universe. Hydrogen gas burns fast and hot—making it a good fuel. Hydrogen is already used to fuel the largest space rockets. Burning hydrogen fuel produces no greenhouse gases or other polluting chemicals that fossil fuels produce—just a puff of steam. The supply of hydrogen will never run out, and it can be made from plain water. However, before hydrogen becomes the fuel of the future, scientists must overcome several technical challenges.

OUR HYDROGEN FUTURE

Hydrogen is the perfect fuel of the future. Hydrogen can be compressed into a liquid and pumped like gasoline to power automobiles and aircraft. Hydrogen is also used in fuel cells that generate electricity. In a fuel cell, hydrogen fuel reacts with oxygen to generate electrical power. But before hydrogen can fuel our future, smart grid engineers must develop safer and more efficient ways to produce and handle hydrogen.

Hydrogen colors. Hydrogen is a colorless gas that burns with a bright orange flame. But scientists and engineers often describe the "colors" of hydrogen. The gas is often described as green, gray, blue, pink, or gold. The various colors indicate how the hydrogen is produced and how clean and renewable the source is. Gray hydrogen is made from natural gas or other fossil fuels. This process releases greenhouse gases into the atmosphere. So gray hydrogen is not a clean source of energy.

Blue hydrogen is made in the same way as gray hydrogen, but the gas emissions are captured and made safe before they are released. Blue hydrogen is a clean fuel.

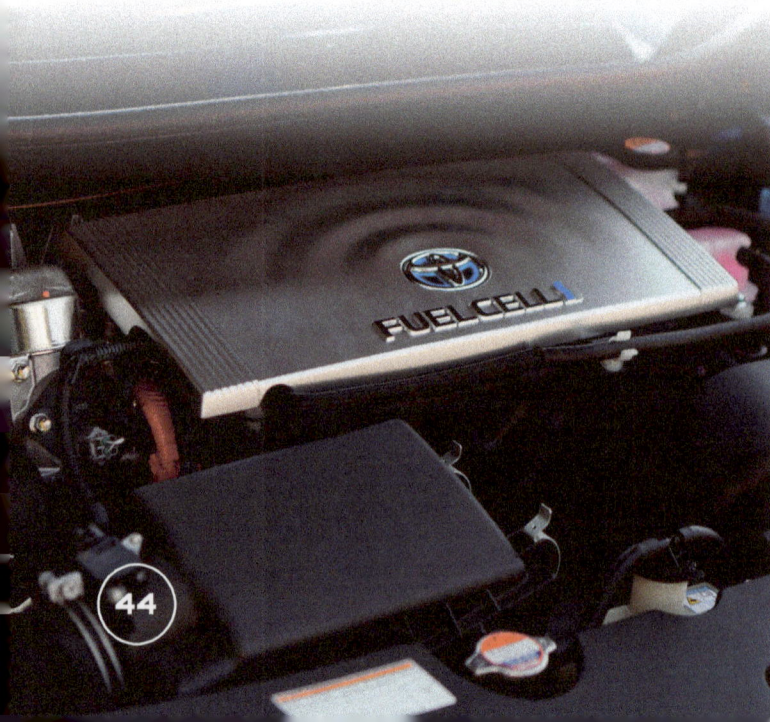

Hydrogen Station

Green hydrogen is made using a process called electrolysis. Here, an electric current is used to split water into hydrogen and oxygen gas. But it can only be classified as green hydrogen if the electricity used to make it comes from a clean and renewable source like solar or wind power.

Pink hydrogen is made by electrolysis powered by electricity from nuclear power plants. Finally, gold hydrogen (also called white hydrogen) is naturally occurring hydrogen gas from underground rock.

Biohydrogen. Researchers are developing methods to produce hydrogen with microbes using the power of sunlight. The method uses microbes called *algae* (simple organisms that live in oceans, lakes, rivers, ponds, and soil) that naturally produce oxygen. Scientists treat the algae so that the microbes produce *biohydrogen* (hydrogen from living things). Large tanks filled with algae could become a clean new way to produce hydrogen fuels.

Solid fuel. Hydrogen is a difficult gas to handle. It leaks and explodes easily, and it must be very cold to turn into a liquid. Researchers are looking at ways to make hydrogen gas easier and safer to handle by storing it inside other materials. Hydrogen atoms are incredibly small and fit into the tiny *pores* (holes) of solid crystals like water soaks into a sponge. Gold and other precious metals are among the best materials to store hydrogen. Researchers are looking for less expensive alternatives!

ENGAGE YOUR READER

Nonfiction writing often includes subject-specific vocabulary terms. Knowing the words related to the topic helps us understand the text itself.

When good readers come upon words they don't know well, they pause and try to figure them out. One tool they use is the glossary, like the one on page 4. Not every word can be defined in a glossary, though!

Authors know this, so they leave clues about words in the text. Next time you encounter a challenging word, stop and look for information about its meaning in the surrounding sentences. Sometimes authors define the term right there in the text! Other times, they'll compare the term to something you may already know. Authors even use punctuation like commas or dashes to clue you in to a word's meaning.

INSTRUCTIONS

1. Consider the list of challenge words and identify where each is used in the text. You can use the Index on page 48 to help you locate each term.

2. Explain how the author described each word. Ask yourself "what is happening in the text?" or "how is this word being used?" as you search for clues about their meanings.

3. Create your own definitions of the words. Don't just copy the dictionary definitions. Instead think about how you would tell a friend what each term means.

4. Add a visual representation for each word. Think about what you could draw that will help you remember what the words mean.

CHALLENGE WORDS

- Grid
- Renewable power source
- Producer
- Decentralized power

- Optimization
- Base load
- Peak demand
- Hydrogen

EXAMPLE

Challenge Word	Page(s)	Author's Description	Personal Definition	Visual Representation
Grid	5, 7	- electrical supply system for a city - first developed 130 years ago, similar technology today - smart grids use AI to efficiently produce cleaner energy	The system used to gather energy from producers, store it, and transmit it to consumers as electricity. Because the current system is so old, it is inefficient and produces a lot of pollution. Smart grids are being developed to produce cleaner energy.	
Renewable Power Source				

INDEX

A

artificial intelligence (AI), 5, 12-13
atomic fission, 30

B

base load, 25-26, 29
biohydrogen, 45
blue hydrogen, 44

C

carbon-capture technology, 27
climate change, 5, 7, 17, 20, 26-27
compressed air energy storage (CAES), 40

D

dam, 10, 36
digital technology, 9
distributed energy resources (DER's), 10-11, 14, 17, 34

E

electric vehicle (EV), 15, 34-35, 39
electricity, 5, 7-15, 17- 20, 22-23, 25-29, 31, 33-36, 39-41, 43-44
Electrostatic Wind Energy Converters (EWICON), 21
energy grid, 5, 7
engineers, 19, 22, 30, 44
expert system, 12

F

fossil fuels, 7, 10, 17, 19, 23, 43-44

G

gas turbines, 22
gasoline, 43-44
generator, electricity, 5, 12, 17
geothermal energy, 5, 17, 28

gold hydrogen, 44
gravitational energy, 33, 38
gravity battery technology, 38
gray hydrogen, 44
green hydrogen, 44
greenhouse gases, 20, 26, 43-44
grid, electricity, 5, 7-15, 17, 21-22, 25, 27, 33-36, 38, 40-41, 43-44

H

home area network (HAN), 15
hydroelectricity, 10, 36
hydrogen, 5, 29-31, 43-45

I

ice battery, 41
ignition, nuclear fusion, 31
infrared, 18
iron batteries, 35
islanding, 11
ITER, 31

K

kerosene, 43

L

lithium-ion battery, 15, 34, 35

M

machine learning, 13
modular nuclear reactor, 29

N

nuclear fusion, 5, 30-31

O

offshore wind farm, 21
optimization, 13

P

peaker power plant, 25
photosynthesis, 19
pink hydrogen, 44
plasma, 30

pollution, 5, 7, 20, 25-27
power plants, 7, 11, 14, 24-30, 38, 44
power tower, 19
pumped storage hydropower, 33, 36-38

R

rechargeable batteries, 33-34
regenerative braking, 39
renewable energy, 5, 7, 16-19, 22, 25, 36, 44

S

scrubber, 26-27
smart grid, 5, 7-15, 17, 22, 25, 27, 33-36, 38, 40, 43-44
smart meter, 8
smokestack, 27
sodium batteries, 35
sodium-cooled reactor, 29
solar canal, 19
solar furnace, 19
solar panels, 5, 9, 10-12, 15, 18-19

T

thermal power plant, 26-27
thermophotovoltaic, 18
tidal lagoon, 22
tidal power, 17, 22
tokamak, 30-31
transformer, 11

V

virtual power plant (VPP), 14
voltage, 11

W

water power, 23
water turbine, 22
wind turbine, 5, 10, 12, 20

www.ingramcontent.com/pod-product-compliance
Lightning Source LLC
Chambersburg PA
CBHW040144200326
41519CB00032B/7592